MOTOR STARTING —AND— CONTROL PRIMER

An introduction to the starting techniques and control of electric motors

STEVEN MCFADYEN

First Edition

Copyright © 2014 Steven McFadyen

All Rights Reserved

Disclaimer

The author and publisher have made every effort to ensure that the information in this book is correct at the time of publication. The author and publisher do not assume, and hereby disclaim, any liability to any party for any loss, damage, or disruption caused by errors or omissions, whether such errors or omissions result from negligence, accident, or any other cause.

Motor Starting and Control Primer

Contents

Preface	v
Introduction to Motor Starting	1
Direct On Line Starting	8
Star-delta Starting	11
Auto-transformer Starting	22
Primary Resistance Starting	27
Rotor Resistance Starting	32
Electronic Soft Start	37
Variable Frequency Drives	44
Summary of Motor Starting Methods	47
How To Calculate Motor Starting Time	51
Useful Motor Technical Information	57
Typical Motor Starting Design Information	59
List of Symbols & Glossary	62

Preface

Electric motors are one of the most common pieces of electrical equipment in service. They are the single largest users of electricity, and it is estimated that they account for between 43% and 46% of total global electricity consumption. Electric motors are found in everything from toys to the largest machines on the planet, and their size can be anything from a few watts to the megawatt range. They are used in humble household appliances, and they also have large industrial applications in areas like mining and petrochemical production.

Motor starting and its associated problems are well-known to many people who have worked on large industrial processes. However, these things are, of course, less familiar to others. This book introduces the reader to common methods of motor starting and control, and it provides a solid grounding in the subject for any student or practicing electrical engineer. The book can be used as an introduction to motor starting for beginners, or as a refresher for those who have spent time away from the field. Having said that, it is the author's hope that even experts will find in it the odd interesting observation.

The first chapter addresses the reasons motor starting is important, and the limitations it places on power supplies. Subsequent chapters cover the most common starting methods: direct on line; star-delta; auto-transformer; primary resistance;

rotor resistance; and electronic soft start. Each starting method is illustrated with a power and control wiring diagram, and the components and their uses are explained. Chapter eight introduces variable frequency drive devices, with an emphasis on how they improve motor starting. The final chapters cover how to estimate motor starting times, some typical technical motor data, and a summary of important formulae. A list of symbols used in the illustrations, and a glossary of terms, are included at the end of the book.

The information contained in this book, coupled with the knowledge gained from studying motor starting techniques, will allow the electrical engineer to design and implement reliable and functional motor starters. The necessary information is presented in a concise and easily accessible manner. The book gives only that which is needed to understand the topic – there are no wordy discussions, and the reader will not be overloaded with irrelevant information. Ideally, this book will be utilised as both a teaching guide and as a handy reference that can be used often.

Introduction to Motor Starting

Motors have been in use for over 100 years, and during that time there has been relatively little change in how they function. The induction motor is by far the most widely used motor in industrial and building applications. As such, this book concentrates primarily on the application of motor starting in connection with induction motors.

Electric motor driving a pump

Induction motors rely on the interaction of magnetic fields to convert electrical power into rotating power. The build-up of magnetic fields and back electro-motive force, or back emf, during the motor starting time introduces transient conditions into the electrical system. These transient events can affect the electrical supply system and other equipment connected to it. The key reasons consideration is given to motor starting are: to limit the transient effects; and to ensure that the motor accelerates the mechanical load correctly.

Motor starting time, starting current, and starting transients

Motor starting time is the period from when the electrical supply is connected to the motor to when the motor accelerates to full speed. The length of the starting period is dependent on the combination of the motor and mechanical load, and it can be anything from a fraction of a second to 30 seconds or longer.

High levels of current are required during the start-up period, and they can have detrimental effects on the electrical supply system and other equipment connected to it. The duration of starting transients depends on the load characteristics and how long it takes the motor to run up to speed.

The figure below illustrates what happens during motor starting. During the starting period a current significantly larger than the motor's normal full load running current is drawn, the magnetic fields within the motor and back emf increase, and the mechanical load accelerates. The start-up current can be as high as five to eight times the full load current.

Introduction to Motor Starting

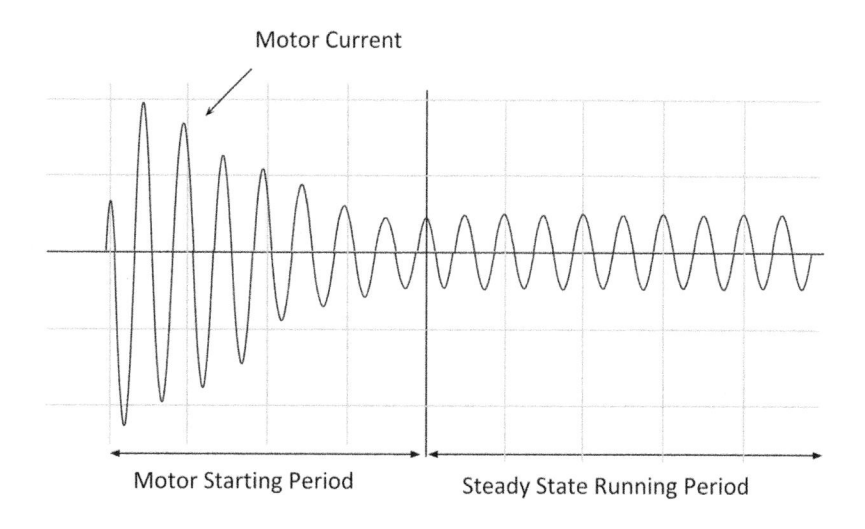

Motor current during starting and running

Electrical systems are designed to cater for the steady state running period conditions. The cables are sized to accommodate the steady state running conditions, and voltage drops across the electrical system are calculated based on the steady state conditions.

During the motor starting period the cables will carry more current than during the steady state running period. System voltage drops will also be much larger during the starting period than during the steady state running period – this becomes particularly apparent when large motors are started, and / or if many motors are started at the same time.

If the voltage drop to the motor itself is too great during the starting period, the motor may be unable to develop sufficient torque to accelerate the mechanical load. In addition, voltage

3

drops within an electrical system may affect other equipment, even to the extent of causing failures.

As the use of motors became widespread, overcoming motor starting problems became a concern for engineers. Over the years, many methods and techniques – each with its own advantages and limitations – have been developed to address the issues around motor starting.

This book addresses some of the more commonly used methods of motor starting.

- Direct On Line (DOL)
- Star-delta
- Auto-transformer
- Primary Resistance
- Rotor Resistance
- Electronic Soft Start

DOL and star-delta are by far the most commonly employed methods of motor starting. However, there have recently been massive strides made in the utilisation of electronics in regulating electrical power to motors, and electronic starting is fast catching up with DOL and star-delta. Additional benefits of electronic starting are covered later in the book, but it is worth noting that advances have enabled a great deal of control over the starting current. These advances can also be utilised to allow the motor to operate with very specific acceleration characteristics.

Reduced voltage during starting

Introduction to Motor Starting

Many of the starting methods addressed in this book rely on a reduction in voltage supplied to the motor during starting. It can be useful to understand, in a very general way, how a reduction in starting voltage will affect the motor. If the voltage is reduced then the current drawn will be reduced proportionally, but the torque will be reduced according to the square of the applied voltage. For example:

If the voltage is reduce by a factor of 2 then the current will be reduced by a factor of 2. However, the starting torque will be reduced by the square of the factor, i.e. by a factor of 4 (2*2 = 4).

The motor starter

A motor starter, or a starting circuit, is a collection of devices that, when functioning together, allows the starting and running of an electric motor. This collection of components is often schematically divided down into the power circuit and control circuit. There may also be an interface to a supervisory control and data acquisition system (SCADA) or other control system.

Power circuit
A power circuit is designed to achieve a few things. It typically contains a means for isolation of the motor system from the power supply, it offers protection against short circuits and overloads, and it provides a means of switching the motor during normal operation.

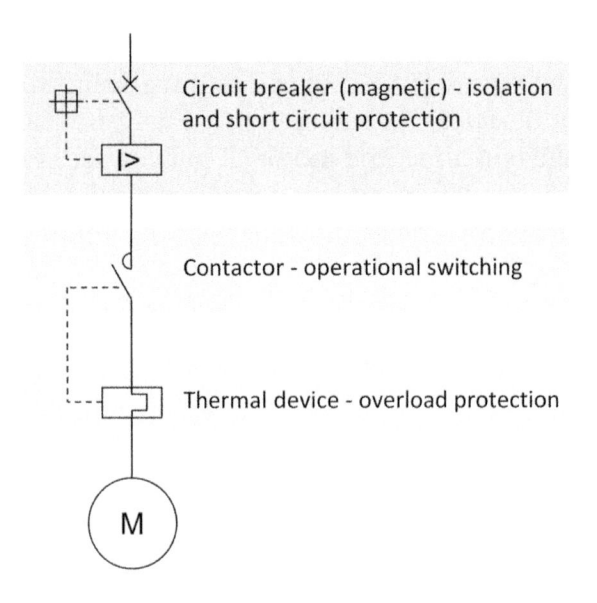

Motor Starting and Control Primer

Circuit breaker (magnetic) - isolation and short circuit protection

Contactor - operational switching

Thermal device - overload protection

Power circuit functions

The above diagram illustrates the main functions of a power circuit. The circuit breaker provides isolation of the motor system and protection against short circuits – older installations may use an isolator and fuse to achieve this functionality. The contactor carries out the operational switching, and it is rated for normal (and overload) currents only. High magnitude short circuit currents cannot be interrupted by the contactor; they should be cleared by a circuit breaker (or fuse). The thermal device provides overload protection.

Any power circuit is typically implemented in one of three ways:

A magnetic circuit breaker provides isolation and short circuit clearance. The contactor provides operational switching, and a separate thermal device provides overload protection.

A thermal-magnetic circuit breaker provides isolation and both short circuit and overload protection. And a separate contactor provides operational switching.

An integrated motor starting solution provides all functions (isolation, short circuit and overload protection, and switching) in a single device.

Most of the examples in this book will be illustrated using the power circuit configuration described in item two.

Control circuit
A control circuit implements the logic associated with starting the motor. There are devices like buttons for starting and stopping (or signal interfaces to control systems), relays for logic and switching the contactors, and timers for controlling transitions.

A variety of control circuits will be illustrated in later chapters, but it is worth noting a few things first. With the increasing complexity of control circuits, and the inevitable issues associated with component failure, many modern motor control circuits are implemented using microprocessor-based relays and controllers. While these devices simplify the physical wiring of circuits, they still need to be programmed to carry out the necessary functional operations. As operational requirements grow, control circuits are becoming more complicated, and the examples offered herein are only the starting point. However, the examples provided will offer a good basis for understanding these more complex arrangements.

Direct On Line Starting

Direct on line (DOL) starting is the simplest and most cost-effective method of motor starting. It is a preferred method of starting because, compared to other starting methods, there is just less that can go wrong. DOL is one of the most commonly employed types of motor starting, and it can be used on very large motors – well into the megawatt (MW) range.

As the name suggests, in DOL starting the motor is simply connected to the electrical supply, and the motor accelerates according to its inherent electrical and mechanical characteristics. DOL starting is the first choice for motor starting if: (1) the power supply can provide the necessary starting currents; (2) voltage drops are not detrimental; and (3) the mechanical load is not adversely affected by high starting torque and acceleration.

Power and control circuiting

The following figure illustrates the power and control circuiting associated with DOL starting.

Direct On Line Starting

Power Circuit

Control Circuit

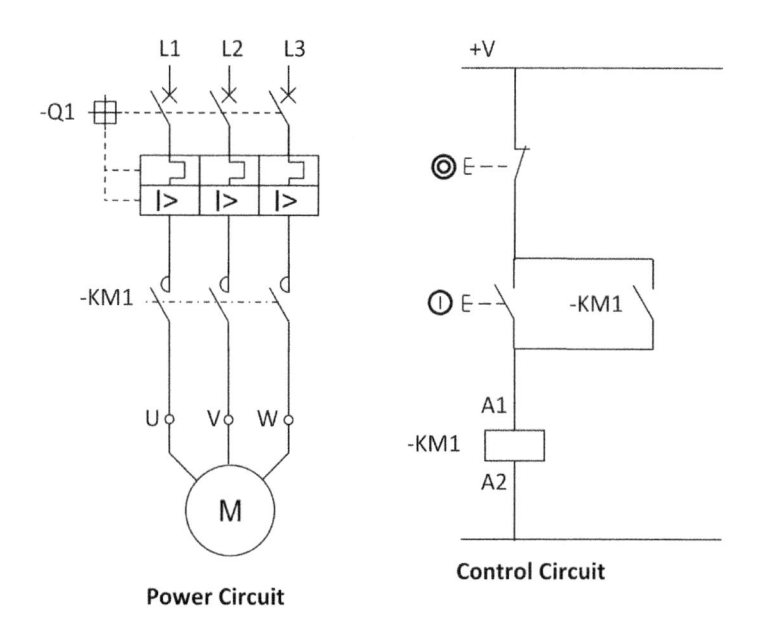

Direct on line starting circuits

As can be seen from the power circuit, closing the switching device (-Q1) connects the motor directly to the electrical supply via a contactor (-KM1). The diagram also illustrates the short circuit and magnetic / thermal overload protective functions.

Push buttons used in motor control circuits are typically momentary, i.e. when they are pressed they will make or break the circuit, and when they are released they revert to their normal state. Due to this, a bypass auxiliary contact of -KM1 is used to maintain power and keep the switching device closed after the start button is released.

The control circuit illustrates a simple push button control for starting and stopping the motor. When the start button is pressed power flows through -KM1 which closes the contactor and starts the motor. Pressing the stop button breaks the power to the circuit. Removing power from -KM1 opens both the contactor and the bypass auxiliary contact.

Although this power / control combination is probably the most basic arrangement for DOL starting, it is the foundation for developing more complex controls.

Direct on line motor starting characteristics

- Available starting current: 100%
- Peak starting current: 4 to 8 In
- Peak starting torque: 100%

Advantages	Disadvantages
Simple	Mechanically Harsh
Low Cost	High Starting Current
High Starting Torque*	High Starting Torque*

* High starting torque is an advantage or disadvantage depending on the application.

Star-delta Starting

Star-delta starting is a reduced-voltage starting method. It is known as Wye-delta starting in the United States. Voltage reduction during star-delta starting is achieved by a physical reconfiguration of the motor windings, as illustrated in the figure below.

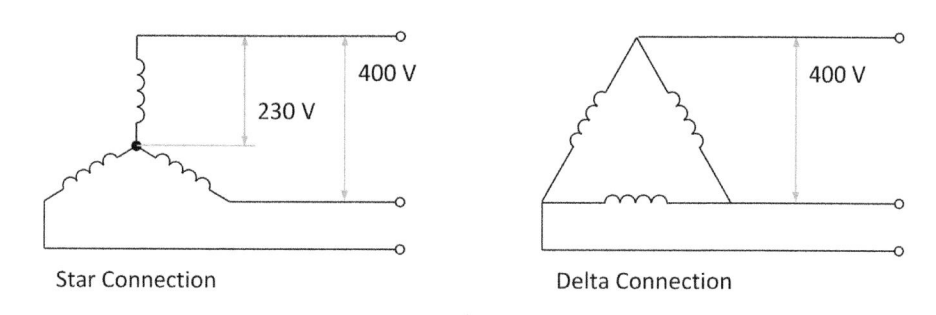

Star Connection   Delta Connection

Star-delta connections

During starting, the motor windings are connected in a star configuration. This reduces the voltage across each winding by the $\sqrt{3}$, and the torque is also reduced by a factor of 3 (see Introduction to Motor Starting). After a period of time, the winding are reconfigured as delta and the motor runs on the normal voltage.

Power and control circuiting

The motor starter – the power and control circuit components – is normally housed in a switchboard assembly some distance from the motor. To enable the changeover from star configuration to delta configuration during starting, all ends of the windings need to be available. This entails the use of a six-core cable between the starter and motor, whereas a DOL starter only requires a three-core cable.

The figures below illustrate the power and control circuits required to enable star-delta starting.

The star-delta power circuit utilises three contactors: -KM1; -KM2; and -KM3. During starting, -KM1 is closed to put the motor into star configuration, and -KM2 is closed to apply power to the motor. At the end of the starting period -KM1 is opened, and then -KM3 is closed to place the motor windings in a delta connection.

Star-delta Starting

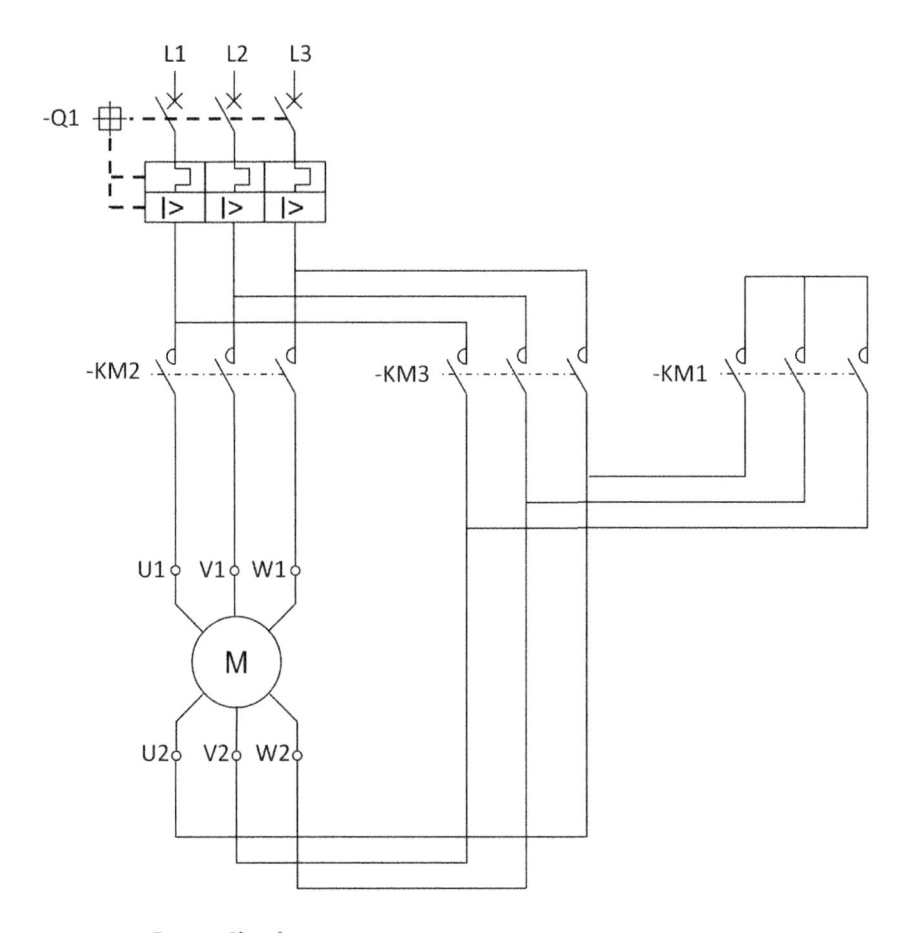

Power Circuit

Star-delta open transition starting circuit (power)

13

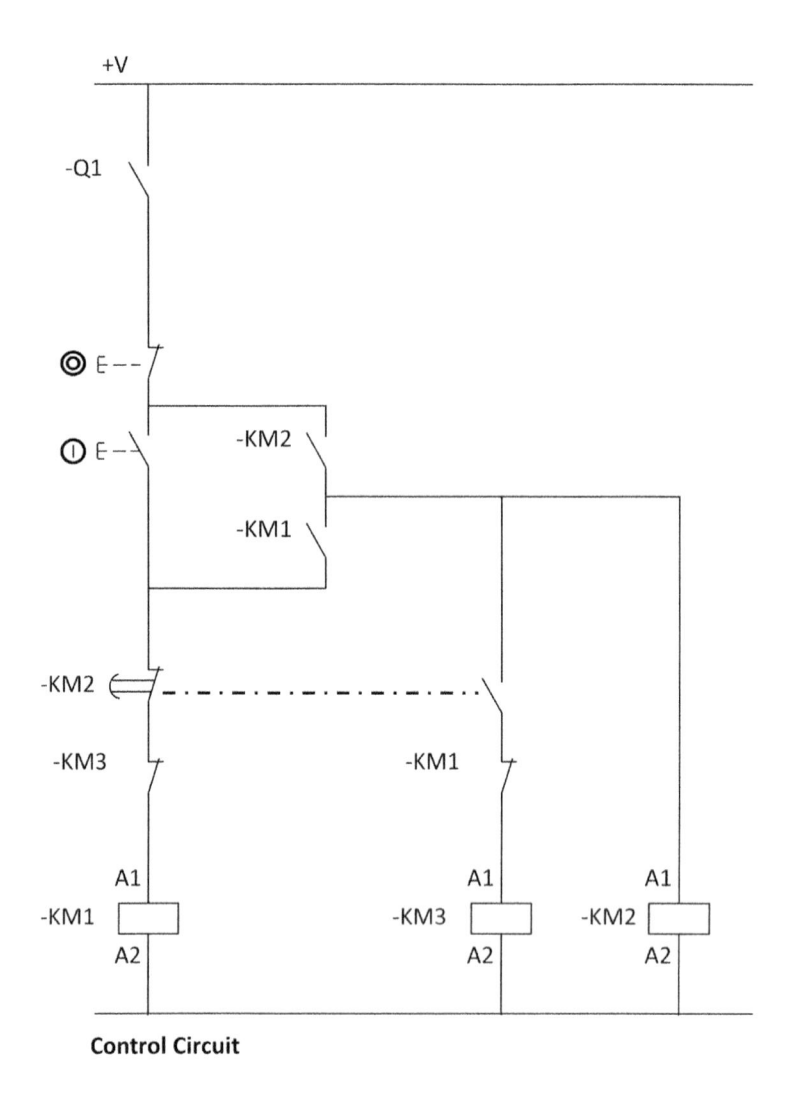

Motor Starting and Control Primer

Control Circuit

Star-delta open transition starting circuit (control)

The star-delta control circuit is more complicated than the control circuit in a DOL starter. Pressing the start button energises -KM1 and closes the power contactor; this enables -KM2 to energise and

14

put the motor into star. After a time delay, the -KM2 contact will switchover and energise -KM3; this opens the star contactor and puts the motor into delta. Pressing the stop button stops the motor by removing all power from -KM1, -KM2, and -KM3.

To avoid shorting the motor, the -KM3 contactor will only close after the -KM1 contactor is opened; this typically occurs after 40 milliseconds. The two contactors are also mechanically interlocked to prevent inadvertent shorting of the motor – should something go wrong with the control wiring.

The changeover time between star and delta depends on the motor and load. The changeover occurs after the motor has run up to near full speed – this is typically 75% to 85% of full load speed. The time delay is normally adjustable, and it can be set to match the starting characteristics of the motor and mechanical load.

Open or closed transition starting

Star-delta starting always requires a change from a star-connected winding to a delta-connected winding. This can be achieved by arranging the power and control circuits in one of two ways – open transition or closed transition.

The method covered in the previous section is open transition starting. In this starting method the power is disconnected from the motor while the windings are reconfigured via external switching. Open transition starting is the easiest to implement in terms of cost and circuitry.

As long as the timing of the changeover from star to delta is good, open transition starting can work well. However, in practice, it is often difficult to set the necessary changeover timing correctly. When the timing is not correct the disconnection and reconnection of the power supply can cause significant voltage and current transients. These transients – voltage drops and disturbances – can be worse than if the motor had been simply DOL started. For this reason, closed transition starting is often specified.

Closed transition starting

In closed transition starting, power is maintained to the motor at all times. This is achieved by introducing resistors to take up the current flow during the winding changeover. In addition, a fourth contactor is required to place the resistors in circuit before opening the star contactor, and then to remove the resistors once the delta contactor is closed.

The following diagrams show typical closed transition power and control circuits for a star-delta starter. The closed transition control circuit is more complicated than the open transition control circuit. This is because it requires more switching devices in order to carry out resistor switching.

The functioning of the closed transition starter is as follows:

Contactor -KM1 is closed, followed by -KM2. This puts the motor into star connection.

Contactor -KM4 is closed at the end of the star running period. This places the resistors in circuit and the motor in star transition.

Contactor -KM1 then opens. This places the motor in the delta transition phase.

Contactor -KM3 closes, the resistors are shorted, and -K4 can be opened. This puts the motor into delta configuration.

Note: The transition (steps 2 and 3) will typically be 0.1 to 0.5 seconds.

Star-delta Starting

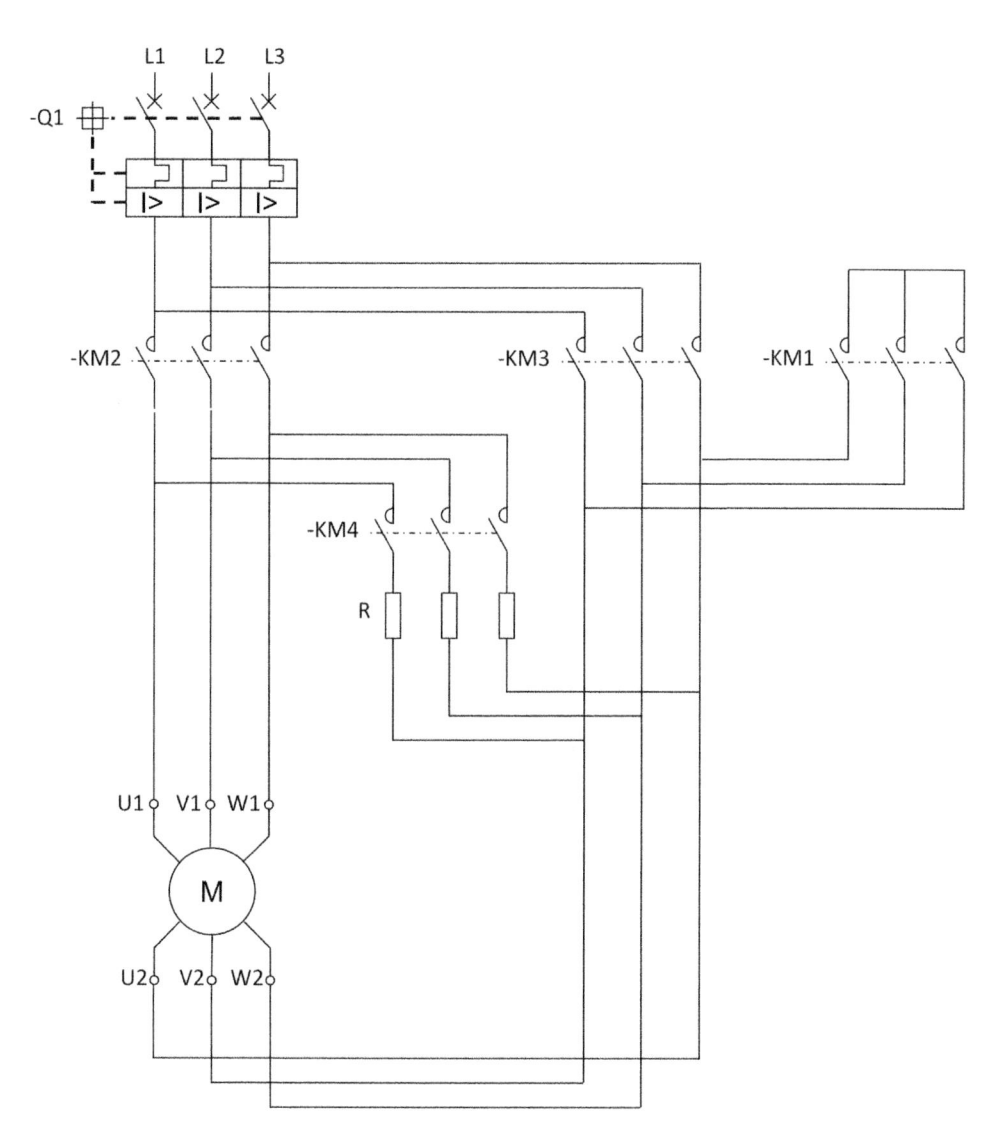

Power Circuit

Motor Starting and Control Primer

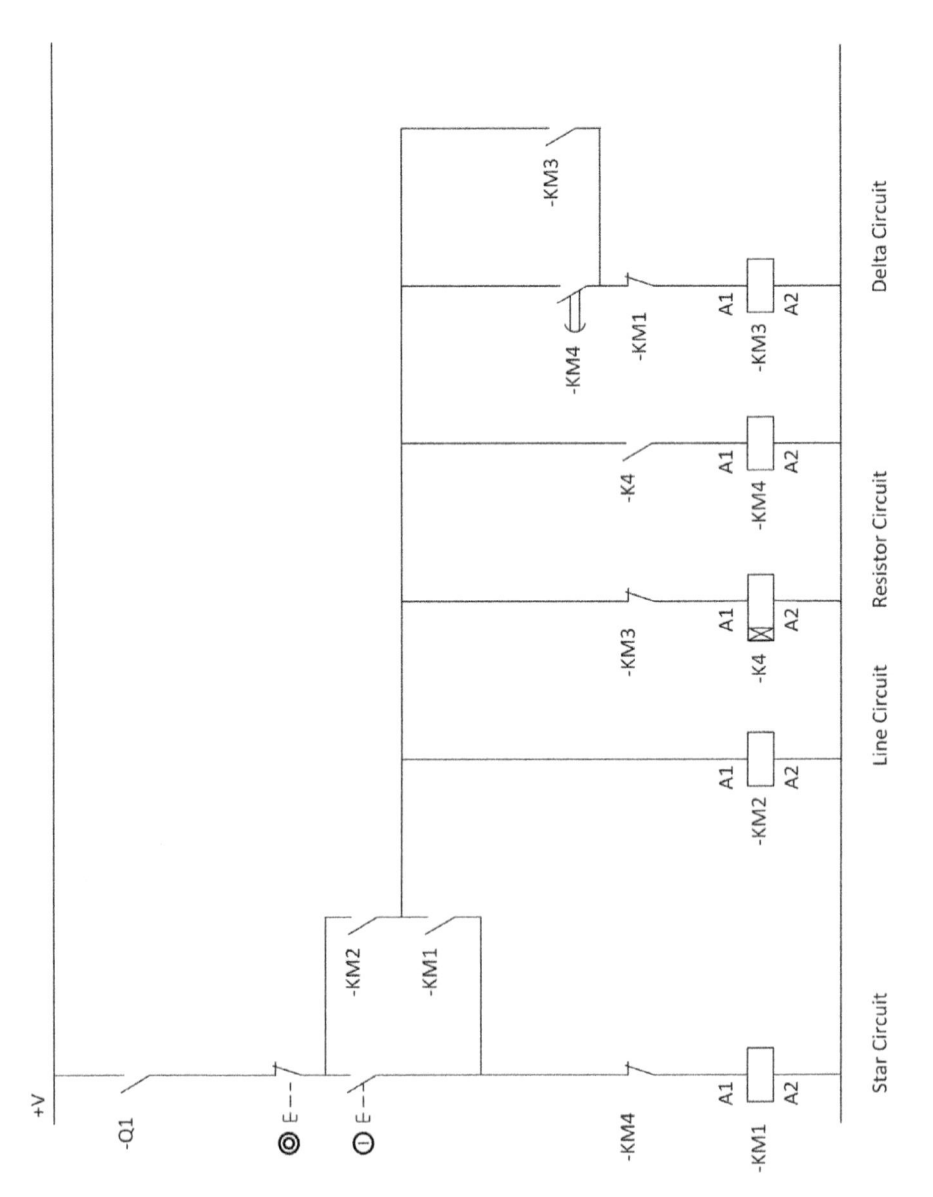

Star-delta closed transition starting circuit (control)

Star-delta Starting

The resistors need to be sized to carry the motor current. Sizing of the resistors is a compromise between wanting to limit the transition current with the star contactor (-KM1) closed (high a value of R), and avoiding an open transition situation when -KM1 opens (low value of R). Optimum resistor selection requires a detailed analysis but, as a rule of thumb, the following equation can be used to determine the size of the resistors:

$$R = \frac{0.33 \times V_{LN}}{I_a}$$

Where

V_{LN} – phase (line to neutral) voltage across the motor

I_a – locked rotor current of the motor

Note: Resistors also needs to be rated to carry the necessary power during the transition, taking into account the likely operating schedule of the motor.

Motor winding connection

The ends of all windings need to be brought out for a star-delta connection. They are typically taken to a terminal block that is labelled U1 / U2, V1 / V2, and W1 / W2 respectively for each winding.

Note: U, V, and W are the IEC connection designation. For NEMA connections: U1=1; U2=4; V1=2; V2=5; W1=3; and W2=6.

Motor Starting and Control Primer

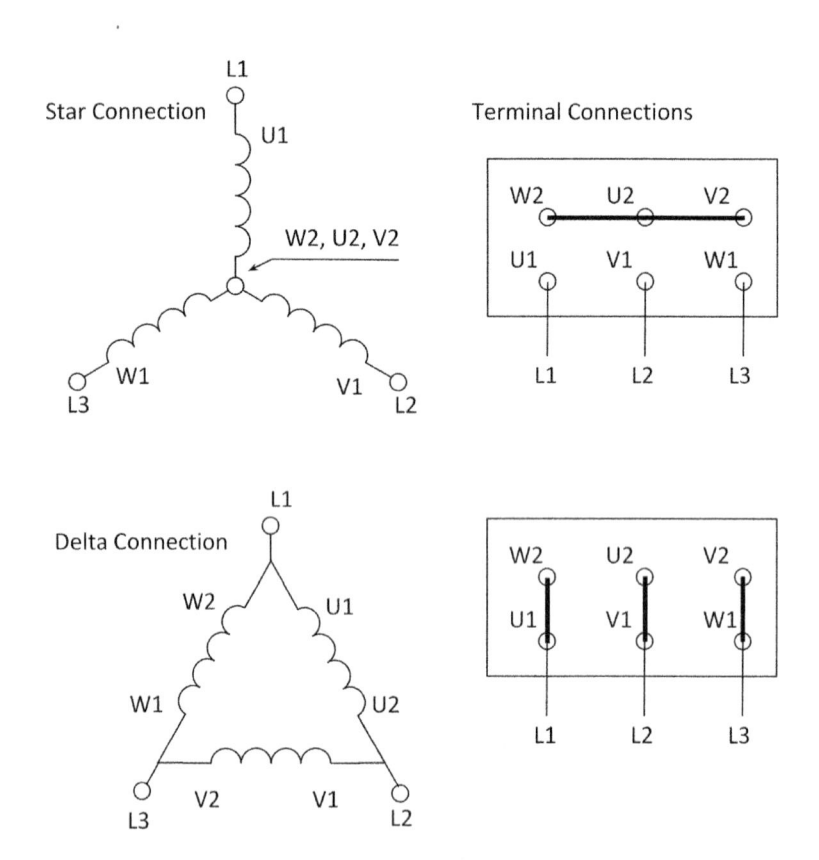

Motor winding connections

The illustration above shows typical motor winding connections. This is not the only way to wire a motor, but it is widely used. There are other ways to wire the motor, they just need to be thought through.

After making the connections as shown, the direction of rotation should be clockwise when viewing the shaft face at the drive end. This should be tested with the motor unloaded. If required, the direction can be changed by swapping two of the phases. The important thing is to make sure both the star and delta are rotating in the same direction, and that it is the desired direction.

Star-delta motor starting characteristics

- Available starting current: 33%

- Peak starting current: 1.3 to 2.6 In

- Peak starting torque: 33%

Advantages	Disadvantages
Simple	Break In Supply – Possible Transients
Low Cost	
Good Torque/Current Performance	Six-terminal Motor Required
Low Starting Torque (depending on application)	Low Starting Torque (depending in application)

Auto-transformer Starting

Auto-transformer starting is, like star-delta starting, a reduced-voltage starting method. Unlike star-delta, which forces a fixed reduction in voltage, the use of an auto-transformer enables almost any range of voltage to be considered during the starting period. The major downside of auto-transformer starting (at least compared to star-delta) is the increased cost.

In auto-transformer starting the voltage can be varied through the starting period due to the use of tappings on the transformer. Reducing the starting voltage reduces the torque according to the square of the applied voltage. For example:

Applying 50% (0.5) of the voltage, i.e. reducing the voltage by 50%, reduces the torque to 25% (0.5*0.5 = 0.25) of the full load current torque. On an 80% tap the torque is reduced to 64% (0.8*0.8 = 0.64) of the full load current torque.

The ability to vary the level of voltage applied during starting allows the starting current to be constrained within limits. It also allows for closer matching of the starting torque to the mechanical requirements.

Auto-transformer Starting

Power and control circuiting

The illustrations below show typical power and control circuits for an auto-transformer starter. In order to illustrate the concept, and to keep it simple, the diagram only shows one transformer tap setting. Using more taps would allow better starting control, but they would also require more contactors and this would expand the control circuit.

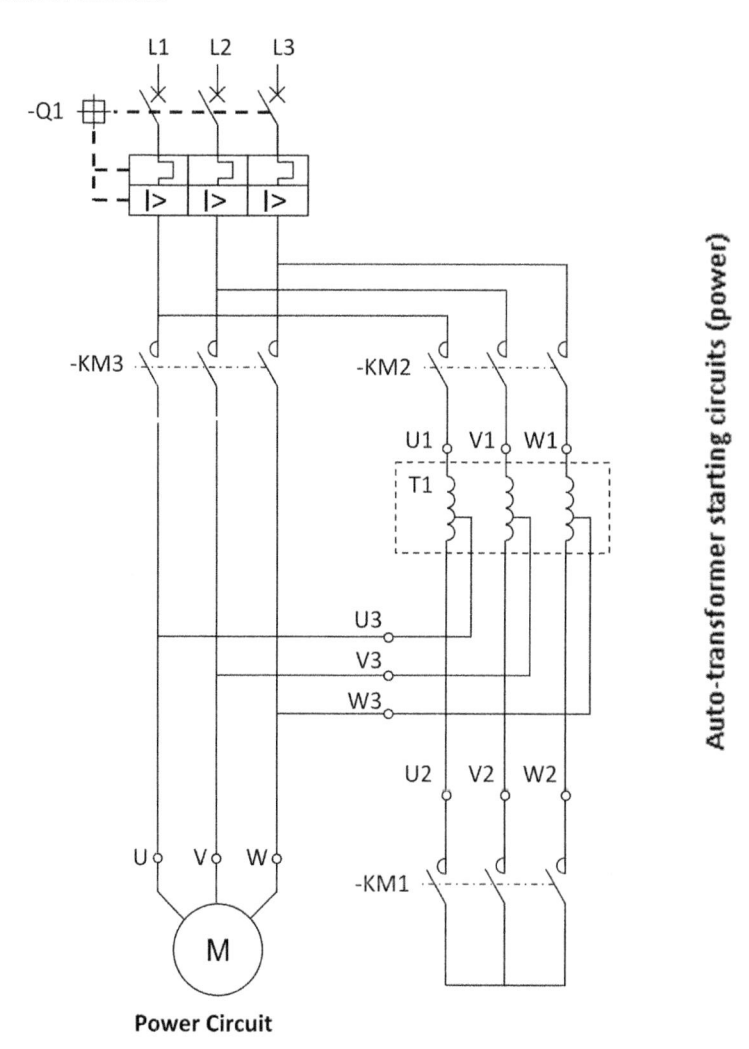

Power Circuit

23

Motor Starting and Control Primer

Control Circuit

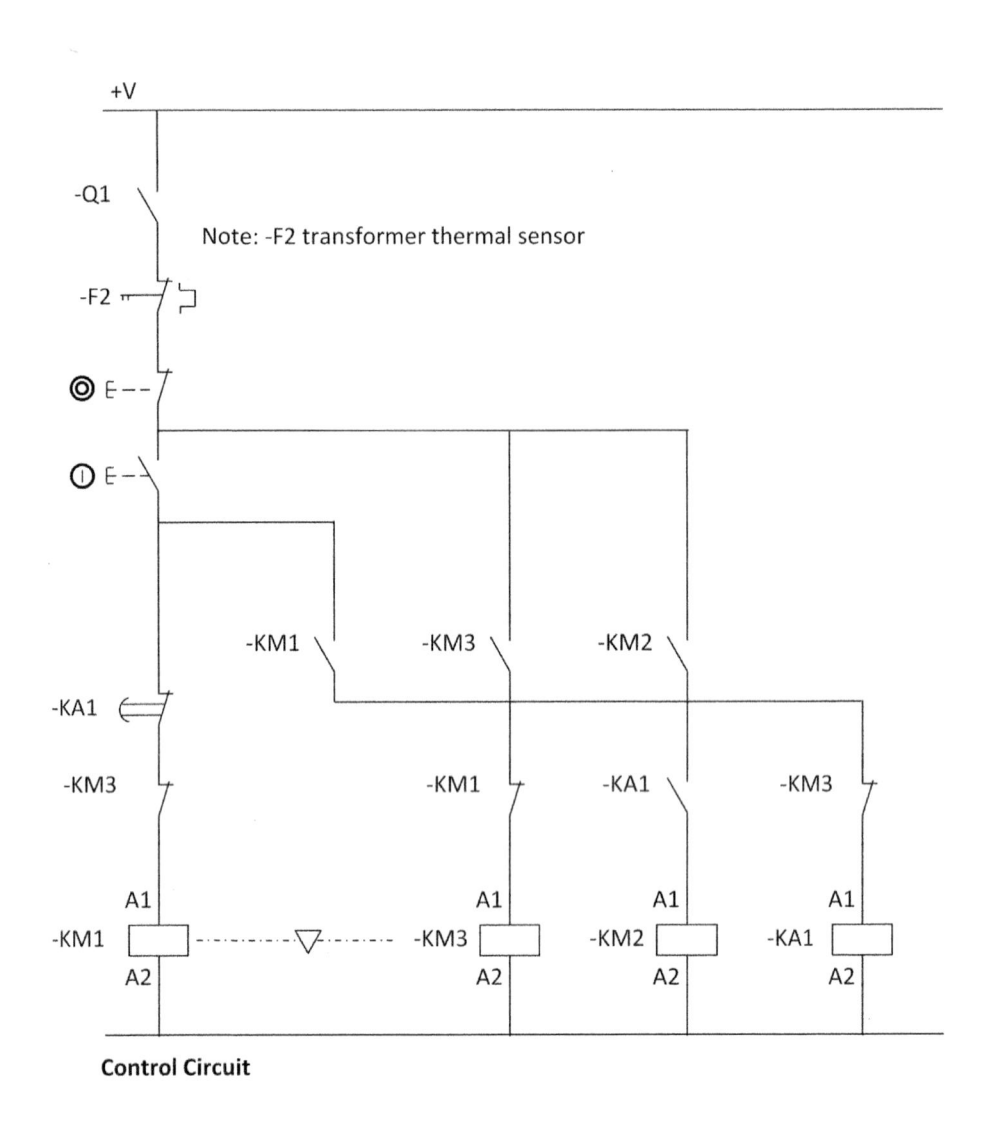

Auto-transformer starting circuits (control)

Auto-transformer Starting

Power circuiting

Auto-transformer starting is typically implemented in three stages:

-KM1 is closed and the transformer is connected in star, then the line contactor -KM2 is closed. This applies a reduced voltage for the initial starting of the motor.

When the motor reaches full speed -KM1, the auto-transformer star contactor, is opened. This momentarily places the transformer in-line with the motor (as an inductance), until step 3 is complete.

The line contactor -KM3 is closed, putting the line voltage on to the motor, and the auto-transformer is isolated by contactor -KM2 opening.

Note: -KM1 and -KM3 are normally mechanically interlocked, thereby preventing both being closed at the same time.

Step 2 ensures that the motor is never disconnected from the power supply, making this a closed transition start. Other control schemes, whereby the motor is momentarily disconnected from the supply while the auto-transformer is removed from the circuit, do exist. However, this type of open transition starting can cause large transient spikes and is not recommended.

Control circuiting

The control circuit in auto-transformer starting is slightly more complicated than the control circuits used in DOL and star-delta starting, but it is easily understandable.

Pressing the start button (-S2) closes -KM1, followed by -KM2 and the sequence timer -KA1. Contactor -KM3 is mechanically interlocked and will not close. Latching contacts keep the relays energised and the motor is started. After a time delay (-KA1), -KM1 will open and -KM3 will close. When -KM3 is closed this will de-energise -KA1 and open -KM2. Pressing the stop button (-S1) will remove power from all devices and stop the motor.

The thermal sensing device (-F2) is used to determine how long the auto-transformer has been energised, and to prevent overheating. It effectively limits the number of starts per hour.

Auto-transformer motor starting characteristics

- Available starting current: 40% / 65% / 80%
- Peak starting current: 1.7 to 4 In
- Peak starting torque: 40% / 65% / 80%

Advantages	Disadvantages
Good Starting Torque / Current Performance Adjustable Starting Parameters No Break in Motor Supply During Starting	Expensive Not Tolerant to Supply Transients

Primary Resistance Starting

Primary resistance starting is concerned with the introduction of resistance into the stator windings to limit the starting current. Resistors inserted into the primary circuit of the motor increase the overall impedance of the circuit and reduce the starting current. In effect, the voltage drop across the resistors gives a reduced voltage to the motor terminals. Two advantages of primary resistance starting over star-delta starting are that the motor windings are not reconfigured, and a six-terminal motor is not required.

The resistors are kept circuit during primary resistance starting, and when full speed is reached they are shorted out – effectively connecting the motor directly to the power supply. The values of the resistors are selected depending on the required limits on starting current and torque.

When the current is reduced in primary resistance starting the torque will also be reduced, possibly more than for a star-delta starter (where the voltage is fixed). In instances where greater control over the current and torque characteristics is required, various resistors can be stepped into or out of circuits during the starting period. It is also possible to use reactors (not resistors) which would allow more options for the control of voltage and torque.

It is worth noting that during primary resistance starting the voltage across the motor is not constant. As the motor accelerates, the current drawn by the motor decreases. This results in a decrease in voltage drop across the resistors, and an increase in the motor winding voltage. The increase in voltage and current in the motor winding mean that the developed torque will increase faster.

Power and control circuiting

The figures below show the power and control circuits of a primary resistance starter.

The functioning of the power circuit is fairly straight forward. Contactor -KM1 closes and the power supply is connected to the motor via the starting resistors. After a time delay, contactor -KM11 closes, shorting out the starting resistors and connecting the motor directly to the power supply.

Functioning of the control circuit begins with the pressing of a momentary start button. This energises -KM1 which is then latched, and power is supplied to the motor via the resistors. After a time delay, a -KM1 delayed contact energises -KM11, shorting out the resistors. Pressing the stop button removes all power and stops the motor.

-F2 is an optional thermal protection device for the starting resistors. It effectively limits the number of starts allowed and prevents overheating of the resistors.

Primary Resistance Starting

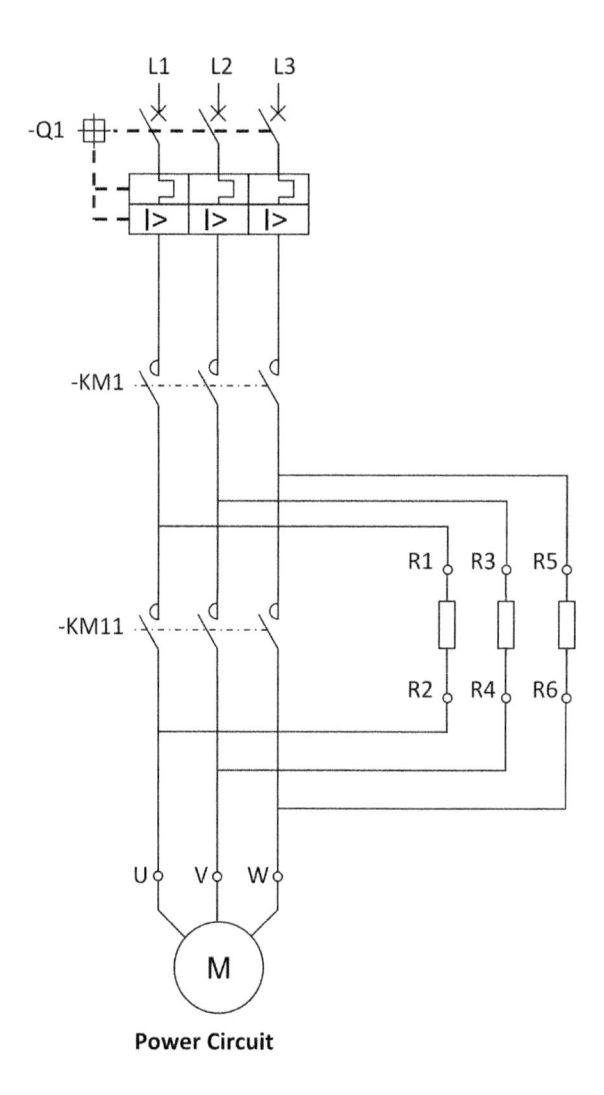

Power Circuit

Primary resistance starting circuits (power)

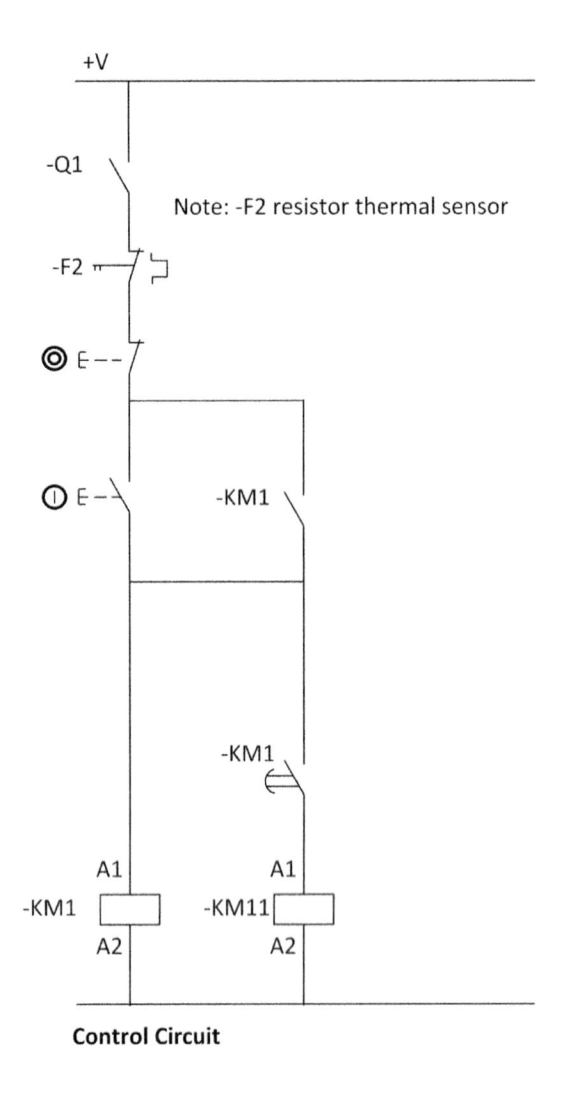

Motor Starting and Control Primer

```
                    +V
                    ┬─────────────────
                    │
         -Q1       \│
                    │
                        Note: -F2 resistor thermal sensor
         -F2  ─┐╱┌─
                │
         ⊚ E ──╱
                    ├──────────────┐
                    │              │
         ① E ──╲    │   -KM1      ╲
                    │              │
                    ├──────────────┤
                    │              │
                    │              │
                    │           -KM1
                    │              ↙
                    │              │
                 A1 │           A1 │
         -KM1  ┌───┐    -KM11 ┌───┐
                 A2 └───┘        A2 └───┘
                    │              │
                    └──────────────┘
```

Control Circuit

Primary resistance starting circuits (control)

30

Calculation of the resistor values is relatively straightforward by using Ohm's law. For example:

Assume a 1.5 kW induction motor, 19 A starting current, system voltage 400/230 V, and required 50% (200 V) voltage drop during starting.

The required resistance R = V/I = 200/19 = 6.63 Ω
The resistor power rating is 126 x 19 = 2.4 kVA (kW)

Primary resistance motor starting characteristics

- Available starting current: 70%

- Peak starting current: 4.5 In

- Peak starting torque: 50%

Advantages	Disadvantages
Adjustable Starting Parameters	Small Reduction in Peak Current
No Break in Motor Supply During Starting	Resistance Bank Required

Rotor Resistance Starting

Rotor resistance starting, like primary resistance starting, involves the addition of resistors into the external circuit. Unlike primary resistance starting, the resistors are connected into the rotor circuit (for resistor connection into the stator circuit see previous chapter). Rotor resistance starting is more expensive and complex than primary resistance starting due to the requirement of slip rings on the rotor and the switching of resistor banks. However, a significant benefit is that the selection of resistors (torque) can be matched to the mechanical requirement of the load.

During rotor resistance starting, banks of resistors are connected into the rotor circuit at the start. As the motor accelerates the resistor banks are switched out on timers until, at full speed, the motor is running directly connected to the power supply.

In rotor resistance starting the torque is nearly proportional to the line current. For example:

A starting torque of 3 times the normal torque will have a starting current of around 3times the normal running current.

Rotor resistance starting is the only motor starting method covered in this book that requires access to the rotor windings. As such, a slip ring motor is required for use with rotor resistance starting – it will not work with a standard squirrel cage motor.

Power and control circuiting

The figures below show typical power and control circuits for a rotor resistance starter.

The power circuit illustrates that contactors -KM11 and -KM12 can be closed in sequence during starting to switch out the resistors. When both contactors are closed the rotor winding is simply connected in its normal running configuration. The number and rating of resistor banks is calculated depending on the mechanical requirements of the load.

The control circuit is relatively straight forward, but bear in mind that this type of starter is controlling switching of both the stator and rotor windings. Pushing the start button energises -KA1, and switches -KM1 which applies power to the stator and at the same time latches -KA1. -KM11 closes after a fixed time delay, switching out resistor bank 1. After another delay, -KM12 closes and resistor bank 2 is switched out, with the motor running fully.

-F2 is a protection device that provides thermal protection for the resistors. It effectively limits the number of starts per hour.

Motor Starting and Control Primer

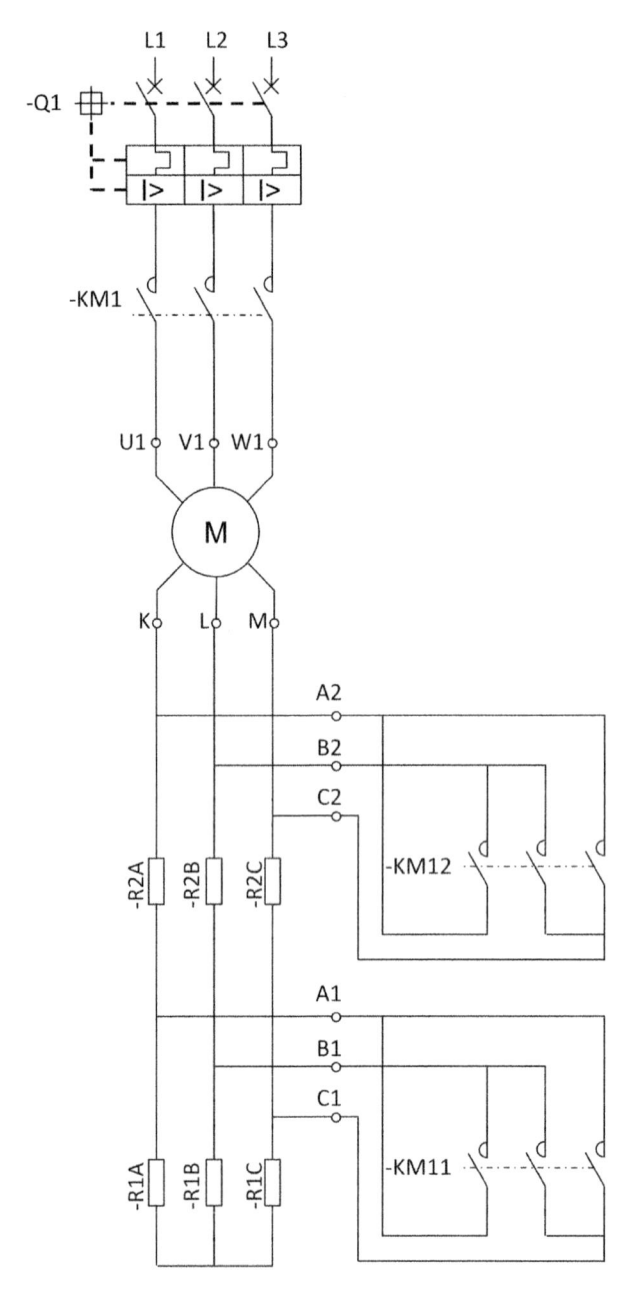

Power Circuit

Rotor resistance starting circuits (power)

Rotor Resistance Starting

Note: -F2 resistor thermal sensor

Control Circuit

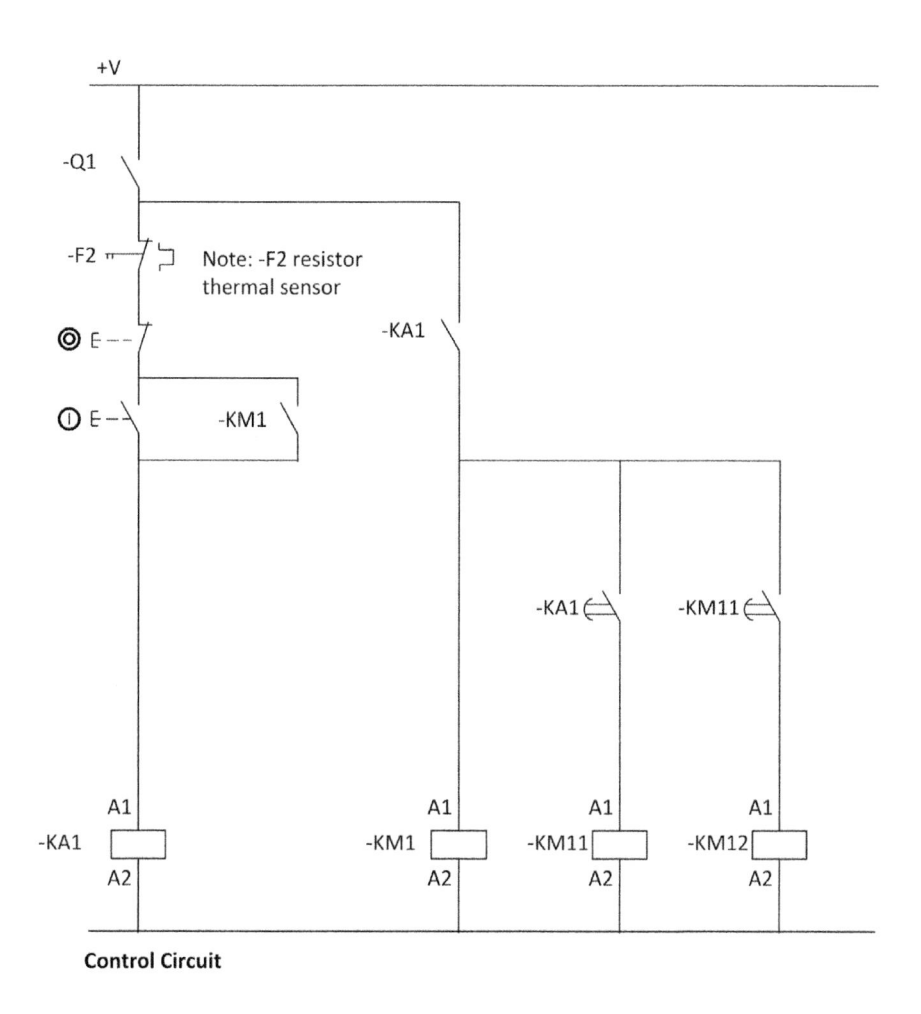

Rotor resistance starting circuits (control)

Rotor resistance motor starting characteristics

- Available starting current: 70%

- Peak starting current: < 2.5 In

- Peak starting torque: ∝ line current

Advantages	Disadvantages
Good Starting Current / Torque Performance Adjustable Settings No Break in Motor Supply During Starting	Expensive Slip Ring Motor Required Resistance Bank Required

Electronic Soft Start

The ongoing development of power electronics continues to revolutionize the way motors are used, and the applications for which they can be utilised. Although books have been written on the application of power electronics to the running and control of motors, that is beyond the scope of this chapter. The chapter will cover the application of electronics in relation to motor starting – and only as an introduction. The discussion will primarily focus on electronic soft starters, but it is important to note that more advanced devices – e.g. variable speed drives – have the same functionality built-in.

Soft starters use a combination of power electronics and electronic control circuitry to slowly increase the voltage on the motor during starting; this ensures a smooth acceleration. Electronic soft starters contain the thyristors (power side) and necessary electronics to control their firing (via user settings). Modern soft starters have a host of features, the most common being: options to set varying start and stop ramps; setting of the initial starting voltage; current limiting control; and thermal overload protection. Soft start units can also be used for stopping the motor by ramping the voltage down. This is particularly useful where sudden loss of driving torque would create mechanical shock on the load.

Motor Starting and Control Primer

Soft starters use thyristors (silicon-controlled rectifiers, or SCRs) to control the energy delivered to the motor. A thyristor is a device that turns on when a pulse is applied to its gate, and it continues to conduct until the current drops to zero – at which time it turns off.

In an AC sine wave current goes to zero each half cycle; this allows the current to be turned off, and it makes it possible to use thyristors to implement soft starting. Controlling the turning on (firing angle) of the thyristors controls the amount of voltage on the motor. If the thyristors are turned on at the start of each half cycle, the full voltage is applied to the motor. If the thyristors are never turned on, then no voltage is applied. If the thyristors are turned on part way through the half cycle, only a proportion of the voltage will be applied to the motor.

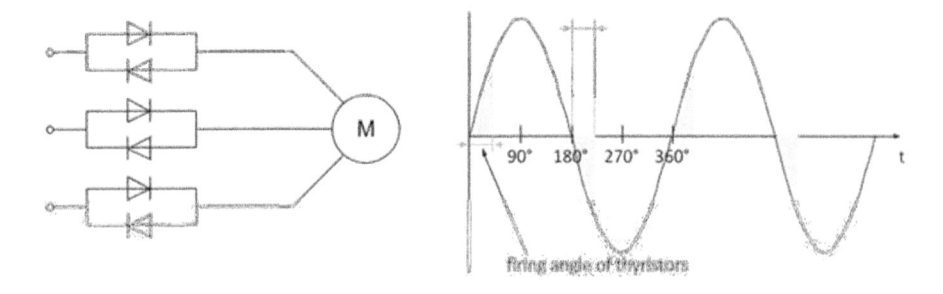

Voltage control by thyristor firing

The sine wave is 360° or 180° for half a wave. The firing angle is just the measure of how far through the cycle the voltage is turned on (and applied to motor). For example:

A firing angle of 45° (180/4), will switch the voltage on a quarter through a half cycle. In the figure above, the firing angle looks to be somewhere between 30° and 40°. A firing angle of 0° will allow the full sine-wave through.

The voltage on the motor will ramp-up during starting by beginning with a large firing delay that is gradually reduced.

As a cautionary note, firing of thyristors other than at zero voltage (current) will create a non-linear load characteristic, generating transients and harmonics. In general application – and given that the transients / harmonics are only present during starting – these are not a problem. However, there could be instances and special situations where these do have an adverse effect on the power system.

The easiest application of electronic soft starters is one unit per motor. To reduce cost, sometimes more than one motor is connected in parallel to a soft starter, or motors are started in sequence by switching the soft start unit between motors. While these methods can be used, care has to be taken to ensure that the units are adequately rated for the required duties.

Power and control circuiting

The figure below illustrates the simplest connection of an electronic soft starter.

In this configuration the soft start unit is simply connected into the circuit and it carries out the necessary functions. This is the simplest implementation, but other implementations will often use a bypass contactor – to switch-out the soft starter when the motor is up to speed – and a line contactor to switch the full starting circuit in / out. Other variations of the control circuit would include cascading start and control for both forward and reverse directions and, invariably, there would also be communications to either control or monitor the functioning of the motor.

Motor Starting and Control Primer

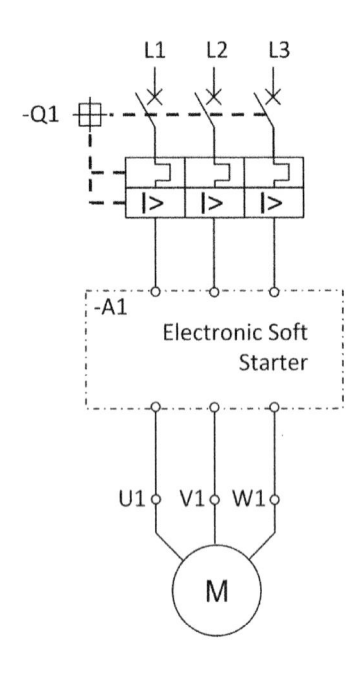

Soft Starter

Features and application

Soft starters have many features not found in traditional starting methods that can be of benefit in some situations. These include:

- **Adaptive Acceleration Control** – the soft starter learns the motor's performance during start and stop, and then adjusts control to optimize performance.
- **Soft Stopping** – slowly bringing the motor to a stop.
- **DC Braking** – injecting DC to reduce motor stopping time.
- **Soft Braking** – changing contactors on starter input, reversing motor direction, and applying braking.

Electronic Soft Start

- **Current Limiting** – limiting starting current to a predefined value (this may lower torque too much and motor may not accelerate).

- **Current Ramping** – increasing the current over a given time to a specified limit.

- **Jog** – allows the motors to be run at reduced speed for some time.

- **Kick Starting** – using a brief high bust of current at the motor start to get it going, then reducing the current.

- **Thermal Modelling** – models the motor's thermal characteristics, allowing performance optimization.

- **Communication** – most industrial protocols will be supported, enabling enhanced control / monitoring.

Line contactors are not strictly necessary, but they can be used to isolate the soft start unit when it is not in use. They should be rated AC3.

If required, **bypass contactors** can be used to minimize heat build up due the to SCRs. They can be rated AC1 as they do not carry starting currents.

Sequence starting can be used to start several motors in sequence, or parallel. Starters need to be rated for full start duty. The necessary additional wiring, contactors, and control relays may not make this an economic option.

Power factor correction, if required, should only be installed on the line side and switched in when the motor is at full speed (AC6 contactors). Capacitors installed on the motor side can cause resonance, increased voltages, and equipment failure.

Motor Starting and Control Primer

Inside delta connection

Inside delta connection is a technique that can be used with six-winding motors. In the configuration below, only half the delta is completed, thus reducing the current the starter is required to carry. The thyristors only carry the phase current, making it possible to start larger rated motors successfully.

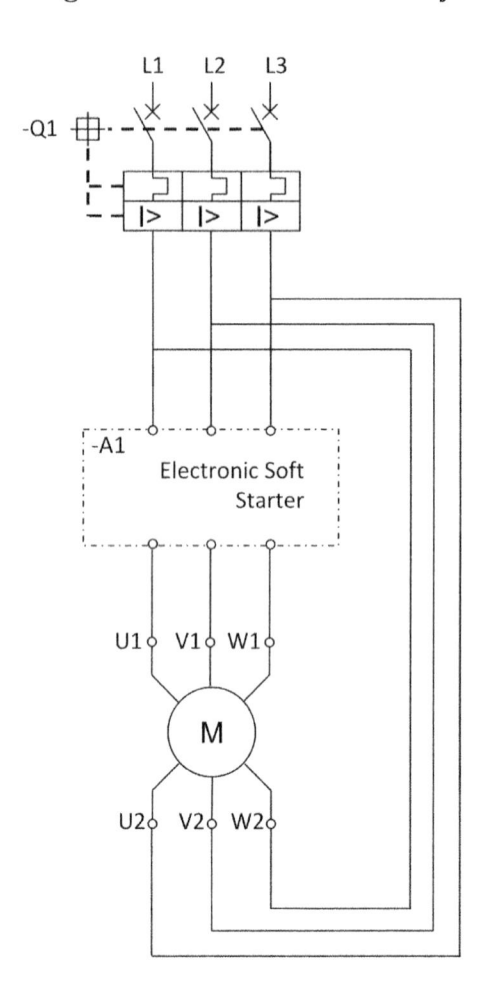

Inside delta connection starter

Only a limited number of soft starters are able to be used in this type of configuration.

Electronic soft start motor starting characteristics

- Available starting current: 25% to 75% (adjustable)
- Peak starting current: 2 to 5 In (adjustable)
- Peak starting torque: 10% to 70% (adjustable)

Advantages	Disadvantages
Fully Adjustable Parameters	More Expensive
Compact	Can Inject Transients Into Supply
Solid State Adaptable to Application	

Variable Frequency Drives

Variable frequency drives (VFDs) are complex devices that offer a multitude of advantages during normal motor operation. However, this chapter will focus on the motor starting aspects of VFDs.

During starting there are a lot of similarities between using a VFD and an electronic soft starter. Anything that can be accomplished with an electronic soft starter can probably be accomplished with a VFD, but VFDs are more costly.

By way of introducing some of the similar functions of VFDs, it is useful to summarize keys functions of electronic soft starters:

- ramp the voltage up during starting to control current and torque;
- varying ramps which can be matched to mechanical requirements;
- setting of initial voltage;
- current limiting control;
- thermal overload protection;
- can be used for controlled stopping.

Additional functionality of VFDs and pulse width modulation

A VFD is designed not only to start the motor, but to control the motor – i.e. to change its speed – during operation. This is achieved by a technique called pulse width modulation (PWM).

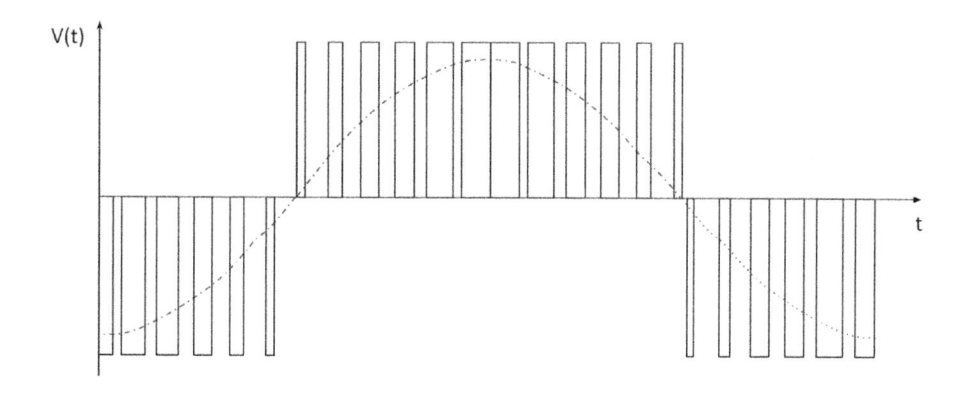

PWM synthesized sinusoidal output

In PWM, insulated gate bipolar transistors (IGBTs) are switched on and off at a high frequency (typically tens of kHz), and this varies the width and frequency of the pulse. This effectively varies the current (rotating magnetic field) to the motor, and therefore its speed. The width of the PWM signal can be varied arbitrarily, and so can the motor speed.

The ability to vary the width and frequency of the pulse has a huge impact on operational performance and what can be achieved with the motor, but it also gives additional options during motor starting. In addition to being used to effectively vary voltage levels, PWM can alter the speed-torque curves during start up (keep in mind that motors are going from zero to operational speed, and no fixed speed during starting is set). Some of the things that can be achieved with a VFD during starting include:

- adjust both frequency and voltage to provide smooth acceleration;
- adjust both frequency and voltage to provide high torque at less than full load current;
- provide almost full torque at zero speed.

There are a couple of precautions to bear in mind in respect of VFDs. Typical motor design may rely on the speed of the motor to drive cooling air across the motor. Running the motor at low speeds for extended periods may reduce the cooling air flow and present a problem. And while it is useful to think of the output as a sine wave (0 to 200 or so Hz), in reality it is a high frequency wave. High frequency noise may be an issue, and there is the possibility of injected harmonics. Measuring instruments need to have the correct frequency response.

Despite any negatives, the amount of motor control that can be achieved with VFDs makes them widely accepted and used in numerous applications. And very importantly, using a VFD for operation will ensure that any starting issues will be much easier to deal with.

Summary of Motor Starting Methods

Electric motors are one of the most common items of electrical equipment in service. Despite the usefulness of electric motors, issues remain around motor starting. Of particular concern is the growth of magnetic fields and back emf during starting that leads to large currents and torque during the starting period. These currents and torque can have negative effects on the electrical system and the mechanical load.

Several starting methods that try to address the problems of starting current and torque are used; each method has its own advantages and disadvantages. The methods are summarized below:

Direct On Line (DOL) is the simplest and most cost effective starting method, and the motor is simply connected to the power supply. This method of starting suffers from high current / torque during starting. Due to its simplicity it is the preferred method of starting, however, there are many instances when its disadvantages make it impractical.

Star-delta is a reduced-voltage starting method. The stator windings are initially connected in a star configuration, and then they are switched to delta when the motor has accelerated. In star, the voltage across each winding is reduced by a factor of $\sqrt{3}$, resulting in a lower starting current and a starting torque of approximately 33% of the full voltage torque.

Auto-transformer is another reduced-voltage starting method whereby an auto-transformer provides the starting voltage. Using an auto-transformer allows the level of voltage – and consequently the current and torque – to be within a wide range of values. With tapped auto-transformers it is also possible to vary the starting characteristics during the run-up period.

Primary Resistance is a starting method that inserts one or more banks of resistors into the stator winding during starting. Voltage drop across the resistors results in a reduced voltage at the motor terminals and improved starting characteristics. As the motor current decreases – and voltage drops across the resistors – the voltage on the winding increases, resulting in a fast increase in torque during starting.

Rotor Resistance motor starting, like primary resistance starting, is controlled by the introduction of resistance banks, but these banks are in the rotor, not the stator. During rotor resistance starting the torque is approximately proportional to the motor current. Selection of the resistor banks can achieve a close match to the required mechanical characteristics during starting.

Electronic Soft Start is a starting method that uses back to back thyristors that are switched to ramp up the voltage during starting (or to ramp the voltage down during stopping). Selection of different ramp characteristics, initial starting voltage, and current limiting functions allow soft starters to match the requirements of the mechanical load, provide smooth acceleration, and ensure that reasonable values of starting current are drawn from the power supply.

Summary of Motor Starting Methods

Variable Frequency Drive (VFD) has many of the same starting functions as an electronic soft start. The driving force in deciding on a variable frequency drive is generally its ability to vary the motor's running speed. Its excellent starting abilities are just a bonus.

The table below summarises the characteristics, advantages, and disadvantages of various starting methods.

Motor Starting and Control Primer

	Direct On Line	Star-delta	Auto-transformer	Primary Resistance	Rotor Resistance	Electronic Soft start
Cost	$	$$	$$$	$$$	$$$	$$$$
Starting Current (xIn)	4 to 8	1.3 to 2.6	1.7 to 4	4.5	< 2.5	2 to 5
Starting Torque	100%	33%	40%/ 65%/ 80%	50%		10% to 70%
Adjustability		+	++	++	++	+++
Typical Load Inertia	Any	Low	Low	High	High	Any
Mechanical Impact	High	Moderate	Moderate	Moderate	Low	Low
Motor Type	Standard	6-terminal	Standard	Standard	Slip ring	Standard
Resistor Bank	No	No	No	Yes	Yes	No

NB: table data is indicative / typical – variations can occur

50

How To Calculate Motor Starting Time

No discussion of motor starting would be complete without considering how to estimate the starting time itself. To accurately estimate the starting time requires some maths, and possibly computer simulation. However, by simplifying various assumptions it is possible to obtain a reasonably good estimate using a simpler calculation.

Starting time is a little complicated

Before looking at the formula that can be applied to get an approximate starting time, it is worth looking at a few of the influences that make accurate calculation difficult.

The first thing to look at is the motor characteristic. The figure below shows a typical motor torque curve with a hypothetical load torque curve (lower curve) superimposed over it.

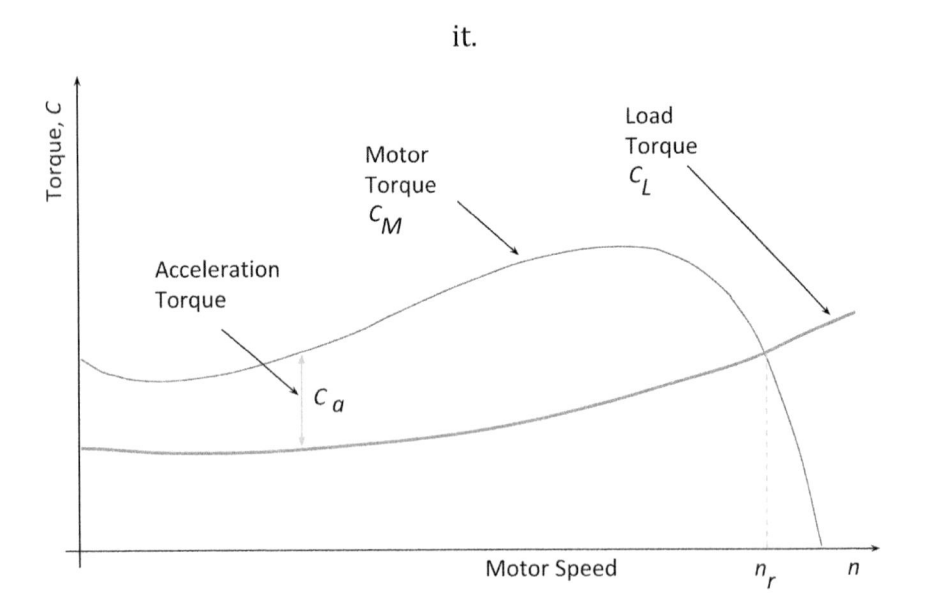

Motor torque speed curve

The torque available to accelerate the motor up to speed is given by the difference between motor torque and load torque. The calculation is as follows:

$$C_a = C_M - C_L$$

Where

C_a – torque to accelerate the motor, N·m
C_M – motor torque, N·m
C_L – load torque, N·m

How To Calculate Motor Starting Time

As the speed increases both the motor and load torques vary. The motor torque characteristic is also a function of the design and construction of the motor, and it can vary significantly for motors of the same rating. Starting methods (see Introduction to Motor Starting) also affect the available motor torque, and can even affect the shape of the curve.

Any torque used for acceleration needs to overcome the inertia of both the motor and the load. By understanding this, and knowing a bit of mechanical engineering, it is possible to derive an equation for the time to accelerate from zero to the running speed as follows:

$$t_a = \int_0^{n_r} \frac{2\pi (J_M + J_L)}{60 (C_M - C_L)}$$

Where

t_a	– time to accelerate to running speed, s
n_r	– motor running speed, rpm
C_M	– motor torque, N·m
C_L	– load torque, N·m
J_M	– inertia of the motor, kg.m²
J_L	– inertia of the load, kg.m²

From the above, by knowing the motor and load inertias, and both the motor and load torque as a function of speed ($C_M(n)$, $C_L(n)$), it is possible to calculate the starting time. While it is possible to do this by solving the equation for an exact solution, in practice, some numerical solution or piecewise approximation would generally be used.

With any complexity on the torque curves or starting arrangement, it is obvious that it is not a trivial matter to calculate the time. For larger, or particularly important motors, the effort of dealing with this complexity would be justifiable. If required, there are software tools available to assist in the calculation.

Starting time – an easier (rough) approximation

By introducing some simplifications, it is possible to have an easier-to-use formula to give an approximation for the starting time.

The first simplification is to use an average value of motor torque.

$$C_M = 0.45 \times (C_S + C_{max})$$

Where

C_S – the inrush torque, N·m
C_{max} – the maximum torque, N·m

Both these figures are available from the motor manufacturer.

For reduced voltages, torque is reduced by the square of the reduction, so it should be possible to adjust the average torque for reduced voltage starting (e.g. star-delta).

How To Calculate Motor Starting Time

The second simplification is to use an adjustment factor K_L to take care of varying load torque C_L due to speed changes.

Typical load factor K_L

- Lift: = 1
- Fans: = 0.33
- Piston Pumps: = 0.5
- Flywheel: = 0

Using the simplifications, the approximate starting time is given as follows:

$$t_a = \frac{2\pi n_r (J_M + J_L)}{60 \times C_{acc}}$$

Where C_{acc} is the effective acceleration torque and is given by:

$$C_{acc} = 0.45 \times (C_S + C_{max}) - K_L \times C_L$$

Motor Starting and Control Primer

An example will show how this works.

A 90 kW motor is used to drive a fan. The motor manufacturer and mechanical engineer have provided information as follows:

- Motor Rated Speed (n_r) – 1500 rpm
- Motor Full Load Speed – 1486 rpm
- Motor Inertia (J_M) – 1.4 kg.m²
- Motor Rated Torque – 549 Nm
- Motor Inrush Torque (C_S) – 1563 Nm
- Motor Maximum Torque (C_{max}) – 1679 Nm
- Load Inertia (J_L) – 30 kg.m²
- Load Torque (C_L) – 620 Nm
- Load Factor (K_L) – 0.33

$$C_{acc} = 0.45 \times (1563 + 1679) - 0.33 \times 620 = 1254.3$$

$$t_a = \frac{2\pi \times 1500 \times (1.4 + 30)}{60 \times 1254.3} = 3.9\,s$$

In summary, while the calculation of motor starting time accurately is not trivial, it is possible make realistic estimates for the most common starting scenarios by using a few simplifications.

Useful Motor Technical Information

Useful induction motor symbols and formulae

Symbols

F – system frequency, Hz

I – motor line current, A

N – motor speed, rpm

n_s – motor synchronous speed, rpm

p – number of motor poles

P_{in} – input power to the motor, W

P_{out} – output (shaft) power delivered by the motor, W

S – motor slip, rpm or %

U_{LL} – motor phase to phase voltage, V

U_{LN} – motor single phase or phase to neutral voltage, V

H – efficiency

Motor Starting and Control Primer

Formulae

$P_{in} = U_{LN} I$ — single-phase motor input power, W

$P_{in} = \sqrt{3} U_{LL} I$ — three-phase motor input power, W

$\eta = \dfrac{P_{out}}{P_{in}} = \dfrac{P_{out}}{U_{LN} I}$ — efficiency, single phase

$\eta = \dfrac{P_{out}}{P_{in}} = \dfrac{P_{out}}{\sqrt{3} U_{LL} I}$ — efficiency, three phase

$n_s = \dfrac{120 f}{P}$ — motor synchronous speed, rpm

$s = n_s - n$ — slip, rpm

$s = \dfrac{n_s - n}{n} \times 100$ — slip, %

$n = n_s \left(1 - \dfrac{s}{100}\right)$ — motor speed (s in %), rpm

Typical Motor Starting Design Information

Single-phase and three-phase average full load motor currents

kW	230 V A	400 V A	500 V A	690 V A
0.06	0.35	0.32	0.16	0.12
0.09	0.52	0.3	0.24	0.17
0.12	0.7	0.44	0.32	0.23
0.18	1	0.6	0.48	0.35
0.25	1.5	0.85	0.68	0.49
0.37	1.9	1.1	0.88	0.64
0.55	2.6	1.5	1.2	0.87
0.75	3.3	1.9	1.5	1.1
1.1	4.7	2.7	2.2	1.6

Motor Starting and Control Primer

kW	230 V A	400 V A	500 V A	690 V A
1.5	6.3	3.6	2.9	2.1
2.2	8.5	4.9	3.9	2.8
3	11.3	6.5	5.2	3.8
4	15	8.5	6.8	4.9
5.5	20	11.5	9.2	6.7
7.5	27	15.5	12.4	8.9
11	38	22	17.6	12.8
15	51	29	23	17
18.5	61	35	28	21
22	72	41	33	24
30	96	55	44	32
37	115	66	53	39
45	140	80	64	47
55	169	97	78	57
75	230	132	106	77
90	278	160	128	93
110	340	195	156	113
132	400	230	184	134

Typical Motor Starting Design Information

kW	230 V A	400 V A	500 V A	690 V A
160	487	280	224	162
200	609	350	280	203
250	748	430	344	250
315	940	540	432	313

NB: Values are only typical and manufacturers' information should be used where possible.

IEC 60974 trip class of thermal protective devices

Class	1.05 Ir	1.2 Ir	1.5 Ir	7.2 Ir
5	t > 2h	t < 2h	t < 2 min	2s < t ≤ 5s
10	t > 2h	t < 2h	t < 4 min	4s < t ≤ 10s
20	t > 2h	t < 2h	t < 8 min	6s < t ≤ 20s
30	t > 2h	t < 2h	t < 12 min	9s < t ≤ 30s

List of Symbols & Glossary

	circuit-breaker	a mechanical switching device, capable of making, carrying and breaking currents under normal circuit conditions and also making, carrying for a specified time and breaking currents under specified abnormal circuit conditions such as those of short circuit
	contactor	a mechanical switching device having only one position of rest, operated otherwise than by hand, capable of making, carrying and breaking currents under normal circuit conditions including operating overload conditions
	make contact	contact which is closed when the relay is in its operate condition and which is open when the relay is in its release condition

List of Symbols & Glossary

Symbol	Term	Definition
⌐	break contact	contact which is open when the measuring relay is in its operate condition and which is closed when the relay is in its reset condition
⇐	contact time delayed	contact in which the action is delayed by some time of receiving the initiation signal
⊃⌐	contact thermal electrical relay	actuating contact of a thermal electrical relay
⊞	trip-free mechanism	a mechanism designed so that disconnection can neither be prevented nor inhibited by a reset mechanism, and so that the contacts can neither be prevented from opening nor be maintained closed against a continuation of the excess temperature or current
E−⌐	push-button momentary	a control switch having an actuator intended to be operated by force exerted by a part of the human body, usually the finger or palm of the hand, which returns automatically to the initial state after operation

Motor Starting and Control Primer

Symbol	Name	Description
(M)	motor	electric machine intended to transform electrical energy into mechanical energy
▢	electrical relay	device designed to produce sudden and predetermined changes in one or more output circuits when certain conditions are fulfilled in the electric input circuits controlling the device **Note:** also used to denote the operating coil of a contactor
I>	overload relay	overcurrent relay which operates when the value of the current exceeds the setting (operating value) of the relay
I>>	short circuit relay	overcurrent relay which operates when the value of the current exceeds the setting (operating value) of the relay
▢	thermal electrical relay	dependent-time measuring relay which is intended to protect an equipment from electrical thermal damage by the measurement of the electric current appearing in the protected equipment
—(■—	plug and socket	plug – connector attached to a cable

List of Symbols & Glossary

		connection	socket – connector attached to an apparatus, or to a constructional element or alike
● ○		junction terminal	**junction** – connection between two or more conductor ends **terminal** – a part of an accessory to which a conductor is attached, providing a reusable connection
ⓘ ◉		- Start - Stop	identification of the function of actuating devices

NB: device descriptions are aligned with the official IEC vocabulary

    Electropedia.org. IEC 60050 – International Electrotechnical Vocabulary – Welcome. [Online] Available from: http://www.electropedia.org/ [Accessed 16 Sep 2013].

65

About the Author

Steven is a chartered electrical engineer with nearly three decades of practical experience working in Europe, Africa, the Middle East, Asia, and Australasia. He's worked on an extensive range of projects, from residential high-rise buildings to transportation systems to mining operations to power stations and petrol plants. His vast experience means that there are very few electrical systems he has not encountered.

Through his work, Steven gained a real appreciation for mentoring and teaching others in his field. In 2002 he started myElectrical.com, an online space for electrical engineers and students. The community members share information, tools, and pose and answer technical and industry-related questions.

It became apparent that Steven's technical expertise is an invaluable support for his colleagues and the myElectrical.com community. Steven's desire to share his knowledge and mastery of electrical engineering with a wider audience has finally led him to the next logical step – publishing his first book: Motor Starting and Control Primer.

Connect with the author online:
http://myelectrical.com/users/steven

Printed in Great Britain
by Amazon